Nikon Z6 III Camera User Guide

Master Your Camera Settings, Autofocus, Video, and Photography Basics—An Easy Step-by-Step User Manual for Beginners, Seniors, and New Z-Mount Creators

Georgette Howard

Table of Contents

11

Introduction

A Camera Worth Mastering—Made Simple

There's something thrilling about holding a new camera for the first time. The soft click of the shutter, the sturdy weight of a quality lens, the way the grip molds into your hand, it feels like the start of something special. But then you turn it on.

And suddenly, it's like stepping into the cockpit of a jet engine.

Menus upon menus. Buttons with cryptic icons. A 400-page Nikon PDF that reads more like a tech manual than a guide for real people. If you've ever felt confused, overwhelmed, or frustrated trying to understand your Nikon Z6 III, you are absolutely not alone. In fact, you're the reason this guide exists.

This Is Not Just Another Camera Manual

Let's be honest: most camera manuals are written by engineers, for engineers. They're cold, clinical, and crammed with technical jargon that can make even the most enthusiastic beginner question whether they made the right choice.

This book is the exact opposite.

Think of it as your personal photography coach. A patient teacher. A friendly guide who knows what it's like to pick up a professional-grade camera and not know where to begin. Whether you're a complete novice, a senior trying to reconnect with your love for photography, or someone simply looking for plain English explanations, you're in the right place.

Why the Nikon Z6 III?

You've chosen a remarkable camera. The Nikon Z6 III is

one of the most powerful and versatile mirrorless cameras on the market today. With its lightning-fast autofocus, breathtaking image quality, stunning 4K and 6K video capabilities, and compact full-frame design, it's trusted by professionals but built for anyone with a vision.

And yet, without the right guidance, it's easy to miss 80% of what this camera can truly do.

This guide will help you unlock every feature and hidden gem of the Nikon Z6 III without the guesswork, without the frustration, and without needing to Google every button.

What You'll Learn Inside This Book

Here's what we'll walk through together:

- How to unbox and set up your camera correctly step-by-step, with real images

- Which buttons do what—clearly labeled, with senior-friendly explanations
- The easiest way to take stunning photos even if you've never touched a camera before
- How to shoot crystal-clear video, without complicated settings
- Real-life shooting scenarios (portraits, landscapes, low-light, kids, travel, and more)
- Troubleshooting help when things go wrong—and they sometimes do
- Pro tricks, accessories, and hidden features to take your creativity to the next level

You won't need to memorize complicated terms or watch 20 YouTube videos to understand your camera. This book will show you the exact settings to use for the situations you care about most with friendly guidance, simple diagrams, and practical tips at every step.

Designed for Beginners & Seniors—With Care

This isn't just a camera guide. It's a confidence builder.

If you've ever told yourself:

- "I'm too old to figure this out."
- "I just want to take great photos without feeling stupid."
- "I'm not tech-savvy enough for a camera like this…"

…this book is here to prove you wrong in the best possible way.

Photography isn't about perfection. It's about seeing. And your eyes, your story, your perspective—they matter. Whether you're capturing a grandchild's laughter, a trip of a lifetime, or the golden hour glow outside your kitchen window, this guide is your bridge between imagination and image.

How to Use This Book

You don't need to read it cover-to-cover. Every chapter is designed to stand on its own, so you can skip to whatever topic you're curious about, be it autofocus, lenses, video, or sharing your photos.

But if you're brand new, I recommend starting at the beginning. We'll walk through everything step by step, as if I were sitting right beside you, guiding your hands and answering your questions along the way.

You'll find:

- Plain English explanations
- Large, easy-to-read formatting
- Checklists, cheat sheets, and pro tips
- Real-world examples not just theory

And most importantly, you'll find yourself growing in **confidence**, **skill**, and **joy** with every page.

Let's Begin This Journey

You don't need to be a professional to use a professional camera.

You just need the right guide.

Let's get started. The Nikon Z6 III is waiting and it's going to change the way you see the world.

Chapter 1

Unboxing & Getting Started Without Confusion

There's something undeniably special about unboxing a brand-new camera. That crisp smell of new gear. The way everything is snugly fitted in its place. The anticipation of all the beautiful photos and videos waiting to be captured.

But then comes the first challenge: what's all this stuff, and what on earth do I do with it?

Don't worry, we'll walk through it all together. This chapter is your simple, stress-free introduction to getting your Nikon Z6 III up and running with complete confidence.

What Comes in the Box: A Visual

Checklist

When you open your Nikon Z6 III kit, here's what you should find inside. Depending on whether you purchased the **body only** or the **lens kit bundle**, some contents may vary.

Note: Always keep the box and all packaging for warranty, resale, or return purposes. If anything is missing, contact your retailer immediately.

Standard Nikon Z6 III Kit Includes:

1. Nikon Z6 III Camera Body

2. Nikon Body Cap (pre-installed to protect sensor)

3. Nikon EN-EL15c Rechargeable Lithium-Ion Battery

4. MH-25a Battery Charger or EH-7P USB-C Charger (region-dependent)

5. Nikon Z-series Neck Strap

6. USB Type-C Cable

7. Hot Shoe Cover

8. User Manual (multilingual)

9. Warranty Card & Registration Paperwork

10. Kit Lens (if included—usually a 24-70mm f/4 S lens)

11. Lens Hood & Lens Caps

12. Viewfinder Eyepiece (sometimes pre-attached)

Quick Tip: Lay everything out on a clean, flat surface and take a quick photo of the contents. It's a good habit for future reference.

Charging the Battery & Inserting Memory Cards (The Right Way)

Step 1: Charge the Battery Fully First

Your Z6 III uses the EN-EL15c battery. Some kits come

with a wall charger (MH-25a) and others with a USB-C charging cable.

To charge using the USB-C method:

- Insert the battery into the camera
- Plug the USB-C cable into the camera and a power adapter
- A charging indicator will light up—let it charge until full

To charge using the MH-25a charger:

- Insert the battery into the charger
- Plug it into a wall outlet
- The orange light means it's charging. Green means it's fully charged

Important: Charge the battery fully before first use to condition it properly.

Step 2: Insert a Memory Card (Gently)

The Nikon Z6 III uses a **CFexpress Type B or XQD** card—professional-grade and incredibly fast. Some bundles also support **UHS-II SD** cards via optional dual-slot adapters.

To insert a memory card:

1. Open the memory card door (right side of the camera)

2. Insert the card with label facing the rear screen

3. Push gently until it clicks into place

4. Close the door securely

Use at least 64GB storage with high read/write speed (300MB/s+ recommended for 4K/6K video)

Understanding Camera Parts (Labeled, With Meaning)

Now let's gently get familiar with what's what.

While you'll find a detailed diagram in the next pages, here's a simple breakdown of the most important parts explained without tech overload:

Back of Camera

- LCD Touchscreen: Use this to frame shots, change settings, and view images

- Viewfinder (EVF): Great for composing in bright light or holding the camera steady

- Menu/Playback Buttons: Access camera settings or review photos

- Multi-selector Joystick: Move focus points or navigate menus

- i Button: Quick settings shortcut menu (a life-saver!)

- AF-ON: Press to focus without shutter (used in back-button focus)

Top of Camera

- **Mode Dial:** Switch between Auto, Manual, Scene modes
- **Shutter Button:** Take the shot (half-press to focus)
- **ISO/WB/Movie Buttons:** Shortcuts to adjust key functions
- **Top Display Panel:** Shows battery, shots left, mode info

Front of Camera

- **Lens Mount:** Where the lens attaches
- **Function Buttons (Fn1, Fn2):** Customize to your needs
- **Grip:** Ergonomically shaped for comfort, especially for seniors

Mode Dial | Movie Record Button | ISO Button | Exposure Compensation Button | Exposure Compensation Button

Playback Button

Function 2 (Fn1) Button

USB Port

Function 2 (Fn2) Button | LCD Screen | Lens Mount | Lens Release Button

Top Display Panel

Function 1 (Fn1) Button

AF-ON Button

Grip

Nikon

Attaching and Removing Lenses

(Beginner-Safe Method)

Handling lenses can be intimidating at first—but don't worry, it's simpler than it looks.

Attaching a Lens

1. Align the marks

To Attach a Lens:

1. Turn off the camera

2. Remove the **body cap** from the camera

3. Remove the **rear cap** from the lens

4. Align the white dot on the lens with the white dot on the camera mount

5. Gently insert the lens and turn clockwise until it clicks

Never force a lens. If it doesn't turn easily, recheck alignment.

To Remove a Lens:

- Press and hold the lens release button (next to the lens mount)

- Turn the lens counterclockwise

- Carefully place the lens with rear cap back on

Bonus Tip: Avoid touching the glass or sensor. Always store lenses face-down on soft surfaces.

Using the Camera Strap (Especially for Seniors)

A good strap isn't just for style, it's safety.

Steps to Attach the Nikon Z Neck Strap:

1. Thread the strap end through the camera lug (metal eyelet)
2. Pass the strap under the strap's plastic fastener
3. Loop back and pull tight
4. Repeat on the other side and adjust length for comfort

Safety First: Always wear your strap during handling,

especially when walking outdoors or switching lenses.

*Optional Upgrade: If the Nikon strap feels rough or tight, consider a **padded shoulder** strap or **hand grip** designed for arthritis or mobility limitations.*

You're Officially Set Up!

That's it! You're now powered up, carded up, and strapped in. And most importantly, you've taken your first step into a world that's about to become more beautiful, meaningful, and creatively alive.

You've got this.

Chapter 2

First-Time Setup — A Friendly Walkthrough

The Nikon Z6 III is a powerful piece of technology but like any finely tuned tool, it shines brightest when properly set up from the start. In this chapter, we're going to walk through every essential first-time setup step together. Think of this as your personal "getting to know you" moment with your new camera.

There's no rush. You don't need to be tech-savvy. Just follow along one step at a time.

Step-by-Step Setup Wizard (Turn It On, Let It Guide You)

When you power on your Z6 III for the first time, it greets you with a simple setup wizard. It's like a quick tour to get your camera ready for the world.

To begin:

1. Insert your charged battery and memory card (from Chapter 1).

2. Turn the **power switch** to the **ON** position.

3. You'll see the **Welcome screen** and be prompted to select your **language**.

Now the camera will walk you through the essential options. Let's go through them with clarity and meaning.

STEP-BY-STEP SETUP WIZARD

Language
- English
- 日本語
- Deutsch   OX
- Francais
- Italiano

SET TIME, DATE & LANGUAGE

Time zone and date

Time zone	UTC:01:00
Date and time	24/04/2024 104(
Date format	DMY
☐ Daylight saing time	☐

OK

BASIC BUTTON NAVIGATION

MENU

Multi selector

Multi selector

DIAL MODES EXPLAINED

SETTINGS TO CHANGE NOW
- Image quality
- Auto ISO
- Focus mode
- Release mode
- Image review

DIAL MODES EXPLAINED

Set Language, Time, Date & Location

These settings may seem minor, but they're important for organizing your photos later and syncing your camera with apps and devices.

1. Language

Scroll with the **multi-selector joystick** or **touchscreen** and choose your preferred language (e.g., English). Press **OK**.

2. Time Zone and Date/Time

Select your region (e.g., GMT +01:00 for Nigeria, UK, etc.), then manually set:

- Year / Month / Day
- Hour / Minutes / AM-PM or 24hr format

Pro Tip: Use the 24-hour format to avoid confusion when

sorting photos.

3. Daylight Saving Time (DST)

Enable this if your location observes DST changes during the year.

This helps ensure your photo metadata is accurate for sorting and archiving.

▣ Basic Button Navigation — What Each Button Really Does

Let's pause here and take a quick beginner-friendly tour of the camera's most used buttons. Don't worry—we'll keep it human.

Top of Camera

- **Shutter Button:** Half-press to focus. Full press to take a photo.

- **ISO Button (Top-right):** Adjusts image sensitivity.

- **Mode Dial (left):** Switch between Auto, Manual, etc. (we'll cover this in a second).

- **Movie Record Button (small red dot):** Instantly start video recording.

- **Display Panel:** Shows remaining shots, battery level, and active mode.

Back of Camera

- **Menu Button:** Opens the full settings menu.

- **i Button:** Opens the shortcut menu for frequently used settings.

- **Playback Button (▶):** View photos you've taken.

- **Multi-selector (Joystick):** Navigate menus or move focus point.

- **Info Button:** Change how information appears on the screen.

- **Zoom In/Out Buttons:** Review and zoom into your photos.

Confidence Tip: Don't feel like you need to memorize all of this. You'll naturally learn these as we go.

Dial Modes Explained — What Each Mode Is Actually For

The **Mode Dial** is your creative command center. Let's make sense of those letters and icons.

Mode	What It Means (Real-World Use)
Auto (Green Camera Icon)	Fully automatic. Camera does the thinking. Great for beginners.
P (Program Mode)	Auto exposure with control over ISO and white balance. Easy to use.

S (Shutter Priority)	You control shutter speed (good for motion), camera picks rest.
A (Aperture Priority)	You choose aperture (depth of field), camera handles other settings.
M (Manual)	You control everything. Best for advanced users.
Scene Modes (SCN)	Portrait, Landscape, Sports, Night—preset for specific situations.
Movie Mode	Switches camera for video recording. Enables video settings.
U1/U2/U3 (User Settings)	Save custom presets for quick access. Very useful later!

*Simple Rule: Start with **Auto,** then try **Scene Modes**. Once* *comfortable, explore P, S, and A to take control of your*

photography.

Essential Settings You Should Change Immediately

Nikon's factory defaults aren't always the most beginner-friendly. Let's tweak a few key settings so your camera is easier to use from day one.

1. Image Quality:

- Go to **Menu > Photo Shooting Menu > Image Quality**
- Choose **JPEG Fine** for best balance between quality and file size
- (If you're planning to edit photos later, enable **RAW + JPEG**)

2. Auto ISO Sensitivity Control:

- Menu > Photo Shooting Menu > ISO sensitivity settings
- Set **Auto ISO Sensitivity Control: ON**
- Max ISO: 6400 (helps avoid noisy images in low light)

3. Focus Mode:

- Menu > AF > AF-A (Auto-servo AF)
- Enables automatic switching between still and moving subjects

4. Release Mode:

- Menu > Photo Shooting Menu > Release mode
- Set to Single frame (so it doesn't take 20 photos per press)

5. Image Review:

- Menu > Playback Menu > Image Review: ON
- See your photo automatically after taking it

6. Date Stamp (Optional):

- Menu > Photo Shooting > Photo info overlay > Date
- Helpful for organizing printed family photos or travel shots

Optional Advanced Tip: Enable Face/Eye Detection AF for better portraits. (AF > Face/Eye Detection > ON)

You're Officially Configured!

Take a deep breath, you've just completed the full first-time setup of your Nikon Z6 III. You've laid the groundwork. Everything from here on out is about building confidence, exploring creativity, and learning how to bring your vision to life.

If you ever feel stuck or overwhelmed, just flip back to this chapter. You're not expected to know everything overnight. Photography is a journey and you've already taken a big step forward.

Chapter 3

Nikon Z6 III Buttons, Dials & Menus—Demystified

To many new users, a professional camera like the Nikon Z6 III can seem like it's covered in mystery knobs, hidden options, and confusing symbols. But here's the good news: you only need to understand a few of these to take beautiful photos and videos.

This chapter is your personal tour of the Z6 III's buttons, dials, touchscreen, and menus—everything laid out visually and explained in real-world, beginner-first language.

Let's simplify your camera, one button at a time.

Simple Visual Guide to Every

Button, Dial, and Port

You've likely already noticed how button-heavy this camera is but not every button is meant for you to use immediately. Think of them like tools in a toolbox. You only need a few to build a beautiful house.

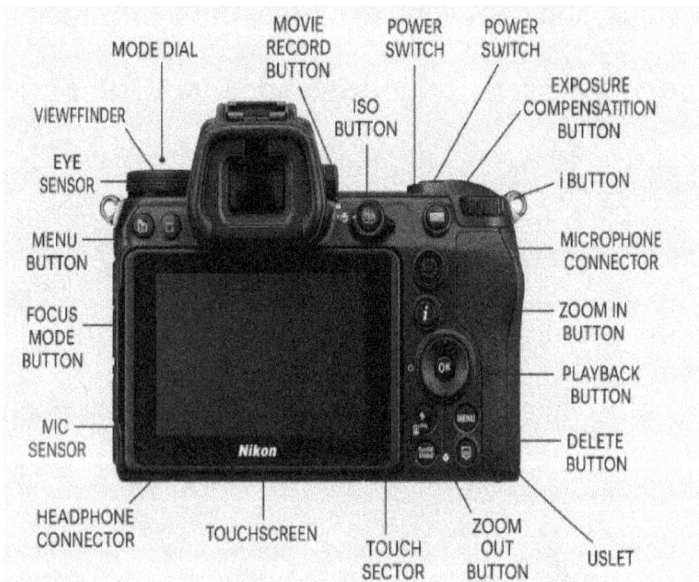

Top of the Camera:

- **Mode Dial:** The wheel that switches between Auto, Manual, and Scene modes.

- **Shutter Button:** Where you press to take a picture. Half-press to focus, full press to shoot.

- **Red Record Button:** Tap this to start or stop video recording.

- **ISO Button:** Changes light sensitivity (we'll guide you on this in later chapters).

- **Exposure Compensation (+/-):** Adjusts brightness of your shot.

- **Top Info Display:** Shows you current settings at a glance (battery, shots left, exposure).

Back of the Camera:

- **Viewfinder (EVF):** Great for outdoor shooting in bright sunlight.

- **LCD Screen:** Touch-sensitive for navigating settings or reviewing shots.

- **i Button:** Your best friend for quick access to key settings.

- **Playback Button (▶):** View photos you've taken.

- **Menu Button:** Opens the full settings menu.

- **Multi-Selector Joystick:** Navigate menus or adjust focus point.

- **Zoom In/Out:** For reviewing photos in detail.

Side of the Camera:

- **USB-C Port:** Charging, firmware updates, file transfer.

- **HDMI Out:** For connecting to external monitors.

- **Mic Input & Headphone Jack:** For video creators.

- **Card Slot Door:** Where you insert your memory card (usually CFexpress/XQD).

Friendly Reminder: You don't need to memorize every button right now. As you shoot, you'll naturally start using the ones that matter to your style.

Understanding the Touchscreen

(Especially for Seniors)

The Z6 III features a responsive 3.2-inch touchscreen that can tilt and flip, useful for both photography and video.

Here's how to make the most of it:

- **Tap to Focus:** Just touch the area you want to focus on.

- **Swipe Through Photos:** Just like a smartphone, swipe left/right in playback.

- **Pinch to Zoom:** Two fingers can zoom in on details while reviewing.

- **Drag Menus:** Scroll through settings or change values with simple gestures.

Senior-Friendly Tip: The screen is very intuitive. If tapping feels unresponsive, adjust sensitivity in Menu > Custom Settings > Controls > Touch Controls.

Customizing Buttons: Simplify Your Workflow

Did you know you can assign specific tasks to several buttons on your Z6 III? This means fewer menus, more convenience.

Here's how to do it:

To Customize Buttons:

1. Press **Menu > Custom Settings Menu (pencil icon).**

2. Scroll to **Controls > Custom control assignment.**

3. Select a button you want to assign (e.g., **Fn1, Fn2, AF-ON**).

4. Choose your preferred function (e.g., ISO, White Balance, My Menu).

Popular Custom Button Ideas:

- **Fn1:** Switch between Focus Modes (AF-S, AF-C)

- **Fn2:** Toggle between RAW and JPEG

- **AF-ON:** Back-button focus (for advanced users)

Bonus: You can save entire setups as User Modes (U1, U2, U3) on the dial for easy recall later!

The i Menu — Your Best Friend for Quick Control

The i Menu is one of the Z6 III's most powerful tools. Think of it as your personal dashboard—customizable and always just one button away.

Where to Find It:

- Press the **i Button** on the back of your camera.

- A grid menu appears on the screen—either in photo or video mode.

Why Use the i Menu?

- Faster than digging through the full menu

- Only shows the most important settings

- You can customize what's displayed

Customize Your i Menu:

1. Menu > Custom Settings > Controls > Customize i Menu

2. Choose whether you want to edit **Photo i Menu or Video i Menu**

3. Assign your preferred shortcuts (Focus Mode, Image Area, ISO, etc.)

Real-Life Shortcut Example:

- Set up your i Menu with:

- White Balance

- Focus Mode

- Silent Shooting

- Picture Control

- Image Quality

50

Now you can tweak your most-used settings in 2 taps, without interrupting your creative flow.

Common Menu Myths Debunked

Let's address some of the biggest fears and misunderstandings users (especially new ones) have about the Nikon menu system:

Myth	Truth
"I'll mess up my camera if I change settings."	Everything is reversible. You can always reset to defaults.
"The menu is too complex for me."	You only need to use a few sections. Ignore the rest.
"RAW is only for professionals."	RAW gives better quality, but JPEG is perfect for

everyday use.

"I must use Manual mode to get good shots."	Not true—Aperture and Auto modes deliver stunning results.
"I need to know every menu option."	Nope. Master 10% of the menu, and you'll unlock 90% of the camera's power.

Encouragement: Don't fear the menu. It's a map, not a maze. We'll continue exploring the key areas together, one purpose-driven step at a time.

What's Next?

Now that you understand your camera's physical controls, touchscreen gestures, and how to customize key functions, you're ready to start shooting like a photographer—not just someone with a fancy camera.

Your camera isn't just smart—it's yours. The more you get to know its buttons and menus, the more it will feel like second nature.

Chapter 4

Lenses for Beginners — Understanding Your Glass

You could own the most advanced camera in the world, but if the lens isn't right for your needs, your photos will suffer. The lens is your camera's eye, its interpreter of light and detail. It's also the most misunderstood piece of gear for new photographers.

In this chapter, we'll clear up the confusion. No jargon. No guesswork. Just a friendly, straight-talking guide to choosing, using, and caring for your Nikon Z6 III lenses.

What Lenses Are Compatible with the Nikon Z6 III?

The Nikon Z6 III uses the **Nikon Z-Mount**—a newer, wider mount system designed for mirrorless cameras. It supports two main categories of lenses:

Native Z-Mount Lenses (Recommended)

These are lenses specifically built for the Z system. They offer:

- Faster autofocus
- Sharper image quality
- Better edge-to-edge performance
- Full compatibility (no adapter needed)

Examples:

- **NIKKOR Z 24-70mm f/4 S** (standard zoom)
- **NIKKOR Z 50mm f/1.8 S** (portrait prime)
- **NIKKOR Z 14-30mm f/4 S** (wide-angle travel)

F-Mount Lenses (DSLR lenses using an adapter)

You can also use older Nikon DSLR lenses via the **Nikon FTZ II Adapter.**

- Pros: Can save money if you already own DSLR lenses

- Cons: May reduce autofocus speed or disable it on some lenses

*Beginner's Tip: Stick with native **Z-mount lenses** unless you have specific F-mount lenses already.*

Prime vs Zoom vs Kit Lenses — What's Best for You?

Understanding lens types is key to capturing the kind of photos you want. Let's break them down simply:

◆ **Prime Lens (Fixed Focal Length)**

- Does not zoom (e.g., 35mm, 50mm)

- Often sharper, brighter (wide apertures like f/1.8)

- Great for portraits, low light, and creamy background blur

Beginner Picks:

- **NIKKOR Z 50mm f/1.8 S** (ideal for portraits, detail shots)

- **NIKKOR Z 35mm f/1.8 S** (versatile walk-around)

◆ Zoom Lens (Variable Focal Length)

- Can zoom in/out (e.g., 24–70mm, 70–200mm)

- Great for travel, events, everyday use

- More flexibility in one lens

Beginner Picks:

- **NIKKOR Z 24–70mm f/4 S** (included in many kits)

- **NIKKOR Z 24–120mm f/4 S** (extended reach for travel)

◆ Kit Lens

- Often bundled with the camera (like 24–70mm f/4)

- Good starting point, but not always the sharpest

- Offers balance between affordability and versatility

Guidance: Start with your kit lens, then explore a prime lens next for better portraits or low-light shots.

PRIME LENS
Fixed focal length
(e.g., 50mm)
Cannot zoom

ZOOM LENS
Variable focal length
(e.g., 24-70mm)
Can zoom in and out

KIT LENS
Typically sold wit
a camera
Affordable optio

Tips for Buying Budget Lenses

Without Regrets

Lenses are an investment—but they don't need to break the bank. Here's how to shop smart:

1. Know Your Purpose First

- Portraits? Go for 50mm f/1.8

- Landscapes? Look for wide-angle like 14–30mm

- Travel? Choose a zoom like 24–120mm

2. Don't Chase Big Aperture Numbers Yet

- f/1.2 sounds exciting but is expensive and hard to master

- f/1.8 or f/2.8 is more than enough for beginners

3. Consider Refurbished or Pre-Owned

- Buy from trusted camera stores (B&H, Adorama, KEH)

- Nikon's official refurbished store offers like-new gear with warranty

4. Skip All-In-One "Superzooms" at First

- Lenses like 18–300mm sacrifice sharpness for convenience
- Better to start with a sharp zoom and add later if needed

5. Always Check Compatibility

- Must be Z-Mount or F-Mount with FTZ adapter
- Check autofocus support before buying third-party options

Using Lens Caps, Cleaning, and Maintenance Tips

Lens Caps (Front & Rear)

- Always keep them on when not shooting

- Prevent scratches, smudges, and dust on glass or mount

Cleaning Routine:

- Use a blower to remove loose dust
- Wipe gently with a microfiber cloth or lens tissue
- For smudges, use lens cleaning solution sparingly

⚠ *Never use a shirt sleeve, paper towel, or household glass cleaner.*

Storage Tips:

- Store in a dry, cool space
- Use silica gel packs to reduce moisture
- Keep lenses upright or in padded bags

Bonus Tip: Consider a UV filter as a protective "shield" for the front element.

The Truth About Adapters & Third-

Party Lenses

Adapters and off-brand lenses can be appealing but they come with trade-offs.

FTZ II Adapter (for F-Mount to Z-Mount)

- Works best with Nikon's newer F-mount lenses that have built-in autofocus motors
- Manual focus may be required on older lenses
- Adds slight bulk to your setup

Third-Party Lens Brands (e.g., Sigma, Tamron, Viltrox)

Pros: Lower cost, good performance (some models)

Cons: Inconsistent autofocus, firmware updates may be needed

Some models may not fully support Eye AF or video focus tracking

Beginner's Guidance: Stick with Nikon Z lenses until you're confident with your style and needs. Then explore third-party lenses selectively.

Wrap-Up: Start Simple, Grow Smart

You don't need a closet full of lenses to be a great photographer. You just need:

- One good lens that matches your subject
- A little knowledge on how to use it
- And the confidence to keep shooting

Let your creativity lead and let your lens follow. Photography isn't about having it all. It's about knowing what you need.

Chapter 5

Auto Modes That Actually Work

If you've ever said, "I just want the camera to do the work for me," you're not alone.

Many beginners and even experienced photographers rely on Auto modes more often than they admit. And there's absolutely nothing wrong with that. The Nikon Z6 III offers several intelligent automatic modes designed to help you capture beautiful photos without needing to learn technical settings right away.

In this chapter, you'll discover the Auto and Scene modes that actually work in real-life scenarios so you can shoot confidently today and grow into manual control when you're ready.

Full Auto vs Scene Modes: Which

Should You Trust?

Let's demystify the options on the Mode Dial and explore which ones are beginner-friendly and which are quietly powerful when used right.

Full Auto (Green Camera Icon)

- The camera does everything for you: exposure, focus, color, ISO, white balance.
- Ideal for: quick family shots, unpredictable lighting, casual use.

Pros:

- Zero learning curve.
- Reliable results in decent lighting.
- Saves mental effort.

Cons:

- Doesn't always know your subject's importance.

- May use slower shutter speeds indoors (leading to motion blur).
- Limited creative control.

Best Use: Outdoor daylight, people, or casual shooting where speed and simplicity matter most.

Scene Modes (SCN Dial Position)

These are like pre-programmed photo assistants built for specific subjects or situations. You pick the scene; the camera picks the best settings.

Common Scene Modes include:

Scene Mode	When to Use It
Portrait	Soft background, flattering skin tones. Perfect for people and pets.
Landscape	Wide scenes, vivid greens and blues. Use for travel,

nature, sunsets.

Close-Up (Macro)	Flowers, food, textures. Best at close distances.
Night Portrait	Low light with flash balanced for natural skin. Ideal for evening events.
Sports	Freezes fast motion. Great for kids playing, wildlife, or active scenes.
Candlelight / Sunset / Snow	Color-sensitive modes for tricky light or color casts.

Hidden Benefit: Scene modes not only adjust camera settings, they also adjust colors, sharpness, and focus behavior to match the subject.

Guide Mode: Nikon's Hidden

Beginner Tool (and Why It Matters)

While many users don't realize it, Nikon has built an intuitive Guide Mode into many of its cameras especially entry-level models.

Although the Z6 III doesn't feature a fully labeled "Guide Mode" like some beginner DSLRs (e.g., D3500), you can create your own guide mode experience by using:

1. The **i Menu** as your quick-access guide
2. **Scene modes** + on-screen suggestions
3. **User Modes (U1/U2/U3)** to save your own "cheat sheet" presets

Your DIY Guide Mode Tip:

Use **U1** for portraits, **U2** for video, and **U3** for landscapes. Save your preferred scene settings into each. It's like having your own preset assistant.

Portrait, Landscape, Close-Up, Night Scene — When to Use What

Here's a simple, real-world cheat sheet to help you choose the right mode at the right time:

Portrait Mode

- Softens background for bokeh (blur), adds skin-friendly color tone.
- Best for: People, pets, indoor photos with natural light.
- Tip: Use a prime lens (like 50mm f/1.8) for stunning results.

Landscape Mode

- Sharp focus from front to back, boosts blue and green tones.
- Best for: Nature, buildings, wide scenes.

- Tip: Hold camera steady or use a tripod for best sharpness.

Close-Up (Macro) Mode

- Enables sharper focus at closer distances.

- Best for: Flowers, food, jewelry, fabrics.

- Tip: Use natural light near a window and steady your hands.

Night Scene / Night Portrait

- Slows shutter speed and adds flash gently to light faces.

- Best for: Evenings, events, parties, cityscapes.

- Tip: Hold still or brace your camera to avoid blur.

Shooting Confidently Without Knowing Manual Settings (Yet)

You don't need to understand ISO, aperture, and shutter

speed to take stunning shots right now. Instead, build confidence by mastering three key things:

1. Composition

- Use the **rule of thirds**: imagine your frame divided into 9 equal parts, and place the subject along those lines.

- Watch your background, keep it simple and uncluttered.

2. Lighting

- Natural light is your best friend. Shoot near windows or outdoors.

- Avoid backlighting unless you want a silhouette.

3. Focus

- Tap the screen to focus if using live view.

- Use Face/Eye AF for portraits, your camera will find eyes automatically.

*Beginner Power Move: Turn on **Auto ISO** + **Scene Mode**, and just watch your results. You'll learn by seeing what works.*

Wrap-Up: Let the Camera Help You—Until You're Ready to Take Over

You don't need to shoot in manual to be a "real photographer." You just need to get started and Auto and Scene modes are powerful tools to help you do exactly that.

Every time you press the shutter in Auto mode, you're gaining:

- Experience in framing and timing
- Familiarity with your camera's responses
- Confidence to grow into manual control

Use the tools you have. Trust your eye. Trust your camera.

Let's build skill by doing—one confident shot at a time.

Chapter 6

Essential Photography Settings—Made Simple

You've probably heard terms like "ISO," "shutter speed," "aperture," or "exposure triangle" tossed around like secret photographer code.

And if it's ever made you feel like this hobby is only for people with a tech background this chapter will change that forever.

You're about to discover that understanding the basics of exposure is easier than you think. Once you grasp these 3 simple elements, your confidence (and photo quality) will skyrocket, no advanced math required.

The Exposure Triangle — Simplified

Photography is the art of capturing light. The three main

settings that control how your camera "sees" light are:

- **ISO:** How sensitive your camera is to light

- **Shutter Speed:** How long your camera lets light in

- Aperture (f-stop): How wide the lens opens to let in light

These three settings form what's called the exposure triangle. Change one, and you affect the others. But don't worry we'll simplify this.

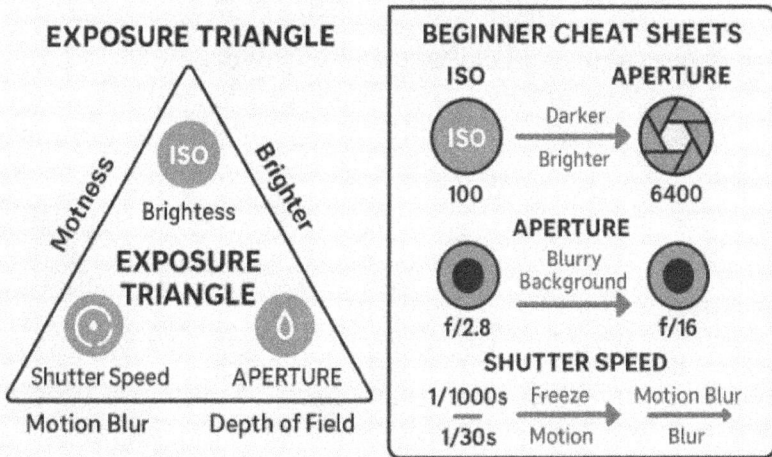

ESSENTIAL PHOTOGRAPHY SETTINGS—
—MADE SIMPLE

EXPOSURE TRIANGLE

ISO

Brightess

Motness / Brighter

EXPOSURE
TRIANGLE

Shutter Speed   APERTURE

Motion Blur   Depth of Field

BEGINNER CHEAT SHEETS

ISO   APERTURE

Darker

ISO

Brighter

100   6400

APERTURE

Blurry
Background

f/2.8   f/16

SHUTTER SPEED

1/1000s   Freeze   Motion Blur

1/30s   Motion   Blur

Think of It Like This:

75

Setting	Controls	Visual Effect	Beginner Tip
ISO	Sensitivity to light	Higher ISO = brighter image, but more grain	Use Auto ISO under 6400
Shutter Speed	Motion blur or freeze	Fast shutter = freezes motion	Use 1/250s for moving kids
Aperture (f/1.8, f/5.6)	Background blur & light	Lower f-number = blurrier background	f/2.8 = dreamy portraits

Shortcut: ISO = Light boost, Shutter = Freeze/Blur, Aperture = Blur background

One-Click Cheat Sheets & Everyday Formulas

Here are simple go-to formulas based on common scenes:

Scene	Formula
Portraits (blurred background)	f/2.8 – 1/250s – ISO Auto
Landscapes (sharp throughout)	f/8 – 1/125s – ISO 100
Sports or kids running	f/4 – 1/1000s – ISO Auto
Indoors (no flash)	f/2.8 – 1/60s – ISO 800–3200
Low Light/Night	f/1.8 – 1/40s – ISO 3200+

Pro Tip: Your Nikon Z6 III can display a live preview—if it looks too dark or blurry, adjust or let Auto ISO help.

Setting White Balance for Real Colors

Ever take a photo indoors and the colors look orange or blue? That's a white balance issue.

White Balance (WB) tells your camera what kind of light is in the room or scene, so it can adjust colors accurately.

Common White Balance Modes:

- **Auto (AWB):** Usually works well—start here!
- **Daylight:** Good for sunny outdoor shots
- **Cloudy:** Warmer tones for overcast days
- **Tungsten (light bulb):** Corrects yellow indoor lighting
- **Fluorescent:** Fixes cool greenish tones
- **Custom Preset:** Can be set using a white sheet of paper under the same light

Beginner Shortcut: Stick with Auto WB unless you notice odd colors, then switch to the matching light type.

Using Auto ISO Effectively (Beginner-Friendly Trick)

Auto ISO lets your camera adjust ISO automatically as lighting changes, while you control the shutter speed or aperture.

To Enable It:

- Go to **Menu > Photo Shooting Menu > ISO Sensitivity Settings**
- Turn **Auto ISO Sensitivity Control** to **ON**
- Set **Max ISO** to **6400**
- Set **Minimum Shutter Speed** to **Auto**

This setup helps you:

- Keep your hands steady without blur

- Shoot in low light without worrying

- Avoid overexposed or grainy images

Set and forget: It's like putting ISO on autopilot so you can focus on framing and focus.

Understanding Metering, Focus Modes & Exposure Compensation

Let's simplify three often-overlooked but powerful settings:

Metering (How Your Camera Measures Light)

Tells your camera which part of the scene it should expose correctly.

Mode	What It Does	When to Use
Matrix (Evaluative)	Measures entire frame	General photography

Center-Weighted	Emphasizes center	Portraits or interviews
Spot Metering	Measures light from a small point	Backlit subjects (like sunset portraits)

Use Matrix for most situations—it's smart and reliable.

Focus Modes (How the Camera Focuses)

Mode	Use Case
AF-S (Single Servo)	For still subjects (portraits, food)
AF-C (Continuous)	For moving subjects (kids, pets, sports)
AF-A (Auto)	Let the camera choose between AF-S and AF-C

Beginner Best Bet: Use AF-A and turn Face/Eye AF ON

for stunning people shots.

Exposure Compensation (+/- Button)

Sometimes Auto Mode makes your photo too dark or too bright. Exposure compensation lets you fix that—without going manual.

- **+1 or +2 = brighter**
- **-1 or -2 = darker**

Use the +/- button on top of your camera, then turn the rear dial to adjust.

Wrap-Up: You Don't Need to Master Everything Today

These settings are here to serve you, not overwhelm you. You don't need to memorize numbers or theory. What matters is this:

- You know what each setting does

- You recognize when your photo needs an adjustment

- You have cheat sheets and presets to guide you

Great photographers aren't born—they're built, one shot at a time. And now, you've got the essential tools to shoot smarter than ever before.

Chapter 7

Autofocus That Actually Finds the Subject

You press the shutter.

The moment's perfect but your subject is blurry, and the background is sharp. Frustrating, right?

If you've ever missed a great photo because the camera focused on the wrong thing or didn't focus at all you're not alone. Autofocus can feel like a guessing game... until you understand how to make your Nikon Z6 III work with you, not against you.

This chapter simplifies autofocus once and for all. You'll learn how to lock focus where you want it, follow fast-moving subjects, and avoid the most common mistakes without ever needing to shoot in manual.

Understanding Nikon's Autofocus System

The Z6 III uses a **sophisticated 273-point hybrid autofocus (AF)** system, combining **contrast detection** and **phase detection** for speed and accuracy. But you don't need to understand the technology—you just need to know which AF mode to use and when.

Here are the three core autofocus modes:

◆ **AF-S (Single-Servo AF)**

- **Focuses once** when you half-press the shutter
- Ideal for **still subjects**: portraits, food, architecture
- If the subject moves after locking focus—it won't adjust

Use AF-S when your subject isn't moving

◆ **AF-C (Continuous-Servo AF)**

- Continuously adjusts focus while holding the shutter button halfway

- Ideal for moving subjects: kids, pets, sports, wildlife

- Tracks subject movement and keeps it sharp

*Use **AF-C** for action, fast movement, or unpredictability*

◆ **AF-A (Auto-Servo AF)**

- **Hybrid mode**: automatically switches between AF-S and AF-C based on subject motion

- Great for **general use** and beginners who want to "set it and forget it"

Best beginner setting: Start with AF-A until you build confidence.

AUTOFOCUS MODES AREA MODES

FOCUS MODES

AF-S
(Single AF)
For still subjects

AF-C
(Continuous AF)
For moving subjects

AF-A
(Auto AF)
Auto switching

AF-A
(Auto AF)
Auto AF

AF AREA MODES

Auto-Area AF
Camera selects
subject

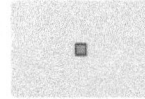

Single-Point AF
One focus point

Dynamic-Area AF
For subjects in
motion

3D-Tracking
Focus follows
subject

Face & Eye Detection AF — When and How to Use It

Your Z6 III has a powerful feature that literally sees like you do, it finds faces and eyes in real-time.

What It Does:

• Detects human (or animal) eyes automatically

- Prioritizes them for razor-sharp portraits

- Tracks faces as they move across the frame

How to Enable It:

- Go to **Menu > Custom Settings Menu > AF**

- Turn **Face/Eye Detection AF** ON

- Choose **Eye Detection Priority** for best results

*For Pet Photography: Use **Animal Eye Detection** Mode instead of Human Eye mode.*

Best Scenarios for Face/Eye AF:

- Portraits (indoor/outdoor)

- Group shots with kids

- Video interviews or vlogs

- Candid street photography

Tip: Combine with AF-C to track eyes as your subject moves.

Tracking Fast-Moving Subjects

(Wildlife, Kids, Pets)

One of the greatest gifts of modern autofocus is subject tracking. This is where your camera "locks on" and follows your subject, so you don't have to keep refocusing manually.

How to Track Subjects:

1. Switch to **AF-C** (Continuous AF)
2. Choose **Wide-Area AF (L)** or **Auto-Area AF**
3. Enable **Subject Tracking** in the AF menu
4. Half-press shutter and the box will follow your subject

Best For: Dogs running at the park, children playing, birds in flight

Tracking Tips:

- Use a **faster shutter speed**: 1/1000s or higher

89

- Use **burst mode (high-speed continuous)** to capture multiple frames
- Pre-focus on the area your subject will move into

Confidence Tip: Subject tracking works best with good lighting and visible contrast.

Common AF Mistakes and How to Fix Them

Mistake	What Happens	Mistake Fix
Using AF-S for a moving subject	Subject becomes blurry	Use AF-C instead
Not enabling Eye Detection	Focus lands on ear or background	Turn Face/Eye AF ON
Using center point only	Misses off-center subjects	Use Dynamic or Auto-Area AF

Subject blends into background	Focus "hunts" and misses	Use Wide-Area AF (L) for better isolation
Focus mode accidentally changed	Camera acts weird	Check AF mode & reset to AF-A or AF-C

Pro Tip: If focus keeps missing, check your lens switch to ensure it's set to AF, not MF (manual focus).

Best Focus Area Modes for Different Scenes

The Z6 III offers multiple AF **area modes**, these tell the camera where to focus in the frame. Here's a cheat sheet for which ones to use:

AF Area Mode	Best For	Why
Auto-Area AF	Beginners	& Camera chooses

	general shooting	best subject
Single Point AF	Still subjects, detail shots	Precise control over what's sharp
Dynamic Area AF	Sports, moving people	Tracks subject when it moves
Wide-Area AF (S/L)	Portraits, pets	Targets larger areas with priority for face/eye
3D Tracking AF	Action, erratic motion	Follows subject across the frame intuitively

*Best Beginner Combo: Use **AF-A** + **Auto-Area AF** + **Face/Eye Detection ON** for most everyday scenes.*

Wrap-Up: Let Your Camera Do the Focusing—You Do the Framing

Your Nikon Z6 III was built to find what matters—eyes, faces, movement, and magic.

You don't need to master autofocus overnight. Start simple. Use Auto-Area and Eye Detection. Let the camera see for you. As you grow, you'll instinctively learn which AF mode fits your moment.

Let the camera chase focus, so you can chase moments. That's the power of understanding autofocus.

Chapter 8

Taking Your First Perfect Shot

You've learned the basics. You've set up your camera. You understand your autofocus. Now comes the moment you've been waiting for:

📷 Taking your first perfect shot.

Not "perfect" by textbook standards—but *your* perfect. The shot that makes you smile. The one that feels just right because you created it.

This chapter gives you the simple steps and techniques to hold your Nikon Z6 III with confidence, frame your subject beautifully, and capture the kind of images you'll want to keep forever.

Holding the Camera for Stability

(Without Shaky Hands)

Even with high-tech image stabilization, nothing replaces proper hand-holding technique—especially when shooting handheld in lower light or using slower shutter speeds.

The Right Way to Hold Your Nikon Z6 III:

1. **Right Hand:** Grips the camera body firmly, with index finger over the shutter button.

2. **Left Hand:** Cradles the lens from underneath—like you're holding a bowl.

3. **Elbows:** Tuck them gently into your sides to create a tripod with your arms.

4. **Stance:** Stand with feet shoulder-width apart for balance.

Bonus Tip for Seniors: If standing gets tiring, try sitting or leaning on a wall to steady yourself. You can also use a neck strap for extra support.

Taking Your First Perfect Shot

Holding the Camara
for Stability

Framing & Composition
Basics

Rule of Thrds   Symmetry

Leading Lines

Holding the Camera
for Stability

Using the Viewfinder vs

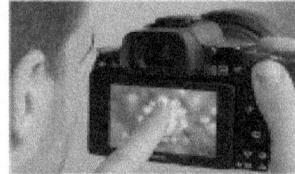

When and How to
Use Burst Mode

LCD Screen

Framing & Composition Basics

(Rule of Thirds, Symmetry &

Leading Lines)

The secret to a "wow" photo often isn't the gear, it's the composition.

Let's explore three timeless techniques to frame your shots beautifully.

1. Rule of Thirds

- Imagine your screen divided into 9 equal parts with two vertical and two horizontal lines.
- Place your subject **on one of the intersections**, not dead center.
- This adds balance and interest.

Try it: Position a person's eyes near the top third line, it instantly feels more natural.

2. Symmetry

Use reflections, doors, hallways, or faces to create pleasing balance on both sides of your frame.

- Best for: Architecture, faces, landscapes with water.

Center your subject only when symmetry is intentional, it makes the shot feel strong and deliberate.

3. Leading Lines

Look for natural lines (roads, fences, shadows) that lead the viewer's eye into the photo toward your subject.

- Use paths, trails, railings, or even your subject's gaze.

Leading lines add depth and pull attention exactly where you want it.

When and How to Use Burst Mode

Burst mode (or Continuous Shooting) takes multiple shots in rapid succession—perfect for moments you don't want to miss.

Best For:

- Kids running or playing

- Jump shots or laughter

- Wildlife and pets

- Sports or action

How to Activate It:

- Press **i Button** or open **Shooting Mode Menu**

- Select **Continuous H (High)** or **Continuous L (Low)**

- Hold down the shutter button and let it rip!

⚠*Caution: Burst mode fills your memory card fast. Review and delete the extras afterward.*

Using the Viewfinder vs LCD Screen

You can compose your photo using:

- The **Electronic Viewfinder (EVF)**

- Or the Rear **LCD Touchscreen**

Which Should You Use?

Viewfinder (EVF)	LCD Screen
Great in bright light	Great for touch focus
Helps steady the camera	Easy for low/high angles
More battery efficient	More intuitive for beginners

se the EVF when outdoors or zooming in. Use the LCD for creative angles and reviewing shots.

Bonus: The Z6 III automatically switches between the two using the eye sensor.

Reviewing Photos, Zooming In & Deleting Mistakes Safely

Reviewing your shots after taking them is essential, it's how you learn, adjust, and get better with every photo.

How to Review:

- Press the **Playback Button (▶)**

- Use the **Multi-Selector** or **Touchscreen** to scroll through images

To Zoom In:

- Press the **Zoom-In button** (+) or pinch on the touchscreen

- Use this to check if your subject is truly sharp (especially the eyes!)

To Delete a Photo:

- Press the **Trash/Delete** Button

- Confirm with **OK**

Smart Habit: Don't delete in frustration. Sometimes slightly "off" photos look great on a computer or in print.

Wrap-Up: Your First Shot Is the Start of Everything

Taking the "perfect shot" isn't about perfection. It's about growth. It's about capturing a moment you saw with your heart and using your hands and camera to keep it.

Whether it's a loved one laughing, a golden evening sky, or just your favorite coffee cup by the window, that photo becomes a memory and you made it happen.

You've got the vision. The tools are in your hands. Now, all you need to do is press the shutter—and trust yourself.

Chapter 9

Shooting Better Videos with the Nikon Z6 III

Still photos are timeless but sometimes, *motion tells the story better.*

Whether you want to capture your grandchild's laugh, a vacation sunset in motion, or record your own video blog or YouTube channel, the Nikon Z6 III is built for it.

With 6K and 4K video, crisp autofocus, and excellent low-light performance, this camera is a video powerhouse—if you know how to unlock its magic.

This chapter walks you through video recording step-by-step: no jargon, no guesswork just the simple settings, accessories, and techniques that will have you recording confidently and creatively in no time.

4K and 6K Video Explained (What It Means for You)

Let's start by understanding the basics of video resolution:

Resolution	What It Means	When to Use It
6K (Raw)	Ultra-detailed, high-quality video (up to 60 fps)	Best for post-production or cropping
4K UHD	Standard high-resolution (up to 120 fps)	Ideal for YouTube, vlogging, interviews
1080p (Full HD)	Lower quality but smaller files	Perfect for everyday family videos or social

*Reality Check: Unless you plan to crop, edit professionally, or slow down motion, **4K is more than enough** for most users.*

For beginners, start with 4K at 30fps (frames per second) for a great balance of quality and file size.

Beginner-Friendly Video Settings (Step-by-Step)

Recording video doesn't require deep technical knowledge. Follow this easy setup for stress-free, high-quality videos:

Step 1: Switch to Movie Mode

- Turn the Mode Dial to the Movie Camera icon
- This activates dedicated video controls

Step 2: Choose Your Frame Rate & Quality

- Go to: Menu > Movie Shooting Menu > Frame Size/Frame Rate
- Recommended setting: 3840x2160 (4K) at 30p

Step 3: Enable Face and Eye Detection

- Menu > Custom Settings > AF > Face/Eye Detection in Video = ON
- Helps keep your subject sharply in focus, even as they move

Step 4: Choose Your AF Mode

- **AF-F (Full-Time AF):** Automatically keeps refocusing (great for vlogs)
- **AF-S (Single):** Focuses once (ideal for still interviews)

Start with AF-F + Face Detection for beginner-friendly ease

How to Focus During Video (Without Blurring)

Autofocus can make or break your video. Here's how to avoid blurry mistakes:

Best Practices:

- Use **Face/Eye** Detection when filming people
- Avoid "focus hunting" by setting **AF-F** and using **Wide-Area AF (L)**
- Tap your subject on the touchscreen to lock focus manually

Pro Tip: For interviews or talking-head videos, use Single AF (AF-S) to lock focus at the start and avoid distracting refocusing.

Recommended Mic Setup for Clean

Audio

Even great video can be ruined by poor sound. The Z6 III has solid internal mics but for truly clear, **professional-quality audio**, use an external microphone.

Best Microphones for Beginners:

1. **Rode VideoMic GO II** (Plug-and-play, no battery required)

2. **Deity V-Mic D4 Duo** (Dual-mic for interviews and ambient sound)

3. **Rode Wireless GO II** (Perfect for vlogs, hands-free filming)

How to Connect:

- Plug into the **Mic Input port** (3.5mm) on the left side of the camera

- Use **Menu > Movie Settings > Audio Levels** to adjust volume manually

Set audio level to peak just below the red zone, this prevents distortion.

Video Modes for Vlogging, Interviews & Home Movies

Let's match the right settings to real-life use cases:

Vlogging (You + Camera)

- **Lens:** Wide (e.g., 24mm–35mm)

- **AF Mode:** AF-F + Eye Detection

- **Frame Rate:** 4K/30p

- **Mic:** Lavalier or on-camera shotgun

- **Stabilization:** Turn on **Electronic VR** for handheld smoothness

Use a tripod or selfie grip to hold camera at arm's length comfortably

Interviews (Stationary Subject)

- **Lens:** 50mm or 85mm (for background blur)

- **AF Mode:** AF-S + Face Detection

- **Frame Rate:** 4K/24p for cinematic look

- **Mic:** External shotgun or lav mic

- **Lighting:** Use window light or soft LED

Lock focus once, don't change camera position mid-recording

Family/Home Videos (Everyday Life)

- **Lens:** Zoom (24–120mm or kit lens)

- **AF Mode:** AF-F

- **Frame Rate:** 1080p/60p (for smoother motion)

- **Mic:** Internal is fine, but external is better

- **Tip:** Keep it casual. Shoot short clips, not long takes.

Bonus Beginner Tips for Better Video

- **Always format your card before recording** (backup photos first)

- Use **Manual Focus** + Magnify to focus precisely on small objects

- Keep your **battery charged**—video drains power fast

- Avoid panning too quickly—it creates a jarring motion

- Use a tripod or gimbal for smoother footage

- Record a short test clip before starting anything important

Wrap-Up: You're Ready to Tell Moving Stories

The Nikon Z6 III isn't just a camera, it's a storyteller. It can record laughter, movement, emotions, and life itself in glorious 4K and beyond.

Whether you're filming a grandchild's birthday, a video journal, or the wind dancing in the trees, this chapter has given you the tools to shoot confidently, clearly, and creatively.

Press record. The story you tell is yours and now you have the gear to tell it well.

Chapter 10

Nikon Z6 III Real-Life Shooting Scenarios

What good is a powerful camera if you don't know what settings to use when life happens?

In this chapter, we'll step away from charts and theory and go hands-on with real-life scenes you'll actually encounter: portraits, vacations, kids, pets, food, night skies, and more.

- Each section gives you:
- The right lens and focus mode
- Exact camera settings to start with
- Simple tricks to improve your results instantly

Let's make the Nikon Z6 III work for you in the real world.

Portraits (Indoors & Outdoors)

Goal: Flattering shots with sharp eyes and soft backgrounds

Setup:

- **Lens:** 50mm f/1.8 or 85mm for dreamy blur
- **Mode:** Aperture Priority (A)
- **Aperture:** f/2.8–f/4 for blurred background
- **ISO:** Auto
- **Focus Mode:** AF-S
- **AF Area Mode:** Eye/Face Detection ON

Light Tip: Use window light indoors. Outdoors, shoot in open shade to avoid harsh shadows.

Landscapes & Travel Photography

Goal: Wide, detailed, colorful scenes with sharp focus

throughout

Setup:

- **Lens:** Wide angle (14–30mm or kit lens at 24mm)

- **Mode:** Aperture Priority (A)

- **Aperture:** f/8–f/11

- **ISO:** 100

- **Focus Mode:** AF-S

- **AF Area Mode:** Single-Point (focus 1/3 into the scene)

Bonus Tip: Use a tripod at sunrise/sunset for long exposures and max sharpness.

Low-Light & Night Photography

Goal: Clear shots with minimal grain, motion blur, or camera shake

Setup:

- **Lens:** Any fast lens (f/2.8 or lower is best)

- **Mode:** Manual (M)

- **Aperture:** f/2.8

- **Shutter Speed:** 1/30s (use tripod if slower)

- **ISO:** 1600–6400

- **Focus Mode:** AF-S

- **AF Assist Lamp:** ON

Pro Tip: Use a remote or 2-sec timer to avoid shake during long exposures.

Action & Kids Playing

Goal: Sharp, in-focus subjects during fast movement

Setup:

- **Lens:** 70–200mm or kit lens at max zoom

- **Mode:** Shutter Priority (S)

- **Shutter Speed:** 1/1000s or faster

- **ISO:** Auto (Max ISO: 6400)

116

- **Focus Mode:** AF-C

- **AF Area Mode:** Dynamic Area or 3D Tracking

Use Burst Mode (High) to capture multiple frames with one press.

Pets & Wildlife

Goal: Sharp eyes, dynamic poses, crisp detail even with motion

Setup:

- **Lens:** Telephoto or zoom (70–300mm)

- **Mode:** Shutter Priority (S)

- **Shutter Speed:** 1/1000s or faster

- **Aperture:** f/4–f/5.6 (for background separation)

- **Focus Mode:** AF-C

- **AF Area Mode:** Animal Eye Detection + Wide Area

Crouch to the animal's eye level—it instantly improves your shot.

Product & Food Photography

Goal: Clean, well-lit, detail-rich images of stationary subjects

Setup:

- **Lens:** 35mm or 50mm prime
- **Mode:** Aperture Priority (A)
- **Aperture:** f/4–f/8 for sharpness
- **ISO:** 100–400
- **White Balance:** Custom or Daylight
- **Focus Mode:** AF-S
- **AF Area Mode:** Single Point (focus on texture)

Use a table lamp with white tissue paper over it for soft DIY lighting.

Vlogging & Home Videos

Goal: Sharp, stable, well-lit personal videos or storytelling

Setup:

- **Lens:** Wide (24mm–35mm)

- **Mode Dial:** Movie Mode

- **Resolution:** 4K/30p

- **AF Mode:** AF-F + Face Detection

- **Mic:** On-camera or wireless lav mic

- **Stabilization:** Enable both IBIS and Electronic VR

Record in small clips, not one long video. It's easier to edit later and less stressful to shoot.

Summary Cheat Sheet

Scene	Mode	Lens	Key Settings

Portraits	A	50mm f/1.8	f/2.8, Eye AF ON
Landscapes	A	14–30mm	f/8, ISO 100
Night Scenes	M	Any low f-stop	f/2.8, ISO 3200
Action/Kids	S	Zoom (70–200mm)	1/1000s, AF-C
Wildlife/Pets	S	300mm+	1/1000s, Animal Eye AF
Food/Product	A	35–50mm	f/5.6, ISO 100
Vlogging	Movie	24mm+	4K/30p, AF-F

Nikon Z6 III
Real-Life Shooting Scenarios

SCENE	LENS	MODE	SETTINGS
Portraits	50mm f1.8	A	f/2.8 ISO Auto Face/Eye AF
Landscapes	14–30mm	A	f/8 ISO 100
Low Light	Any Fast Lens	M	f/2.8 1/30s ISO 3200
Action	70–200mm	S	1/1000s ISO Auto AF-C
Pets & Wildlife	300mm	S	1/1000s ISO Auto AF-C Animal Eye AF
Product	35mm	🎥	f/5.6 ISO 100
Vlogging	24mm	📹	4K/30p AF-F

Wrap-Up: Shoot for the Life You Live

Your Nikon Z6 III is more than capable, it's adaptable. Whether you're capturing stillness or chaos, light or motion, quiet moments or roaring laughter, this camera can handle it all.

The key is knowing which settings serve the moment not the other way around. And now, you have a real-world playbook to guide you.

Shoot boldly. The world doesn't need perfect pictures, it needs your pictures, told with clarity and care.

Chapter 11

Playback, Editing, and Image Review

Taking a photo is just the beginning. The real magic comes when you **review what you captured**, **edit it with a gentle touch**, and **share it with others**.

Whether you're looking to:

- Check if someone blinked,

- Crop a distracting edge,

- Or send your favorite shot to your phone…

…the Nikon Z6 III makes all of it possible right from the camera. And it's easier than you think.

How to View, Zoom, and Rate

Images

After you take a photo, you can immediately review it without switching modes.

To View Images:

- Press the **Playback Button** (▶ icon near the bottom left).

- Use the **Multi-selector joystick** or **touchscreen** to scroll through your shots.

- Press **OK** to access additional options.

To Zoom In or Out:

- Press the **Zoom-In (+)** button to check sharpness—especially the eyes in portraits.

- Press the **Zoom-Out (–)** button to view the full image or grid thumbnails.

- You can also **pinch or swipe** on the touchscreen just like a smartphone.

To Rate or Mark Images:

- While viewing, press the **i Button**, then select **Rating**.

- Assign stars from 1 to 5 to sort or keep track of favorites.

Best Practice Tip: Zoom into the subject's eyes to ensure sharp focus—especially useful for portraits or pet photos.

Using In-Camera Editing (Cropping, Brightness, Filters)

No computer? No problem. Your Nikon Z6 III offers built-in editing tools for quick adjustments on the go.

Available In-Camera Edits:

- Crop
- Straighten
- Brightness/Contrast

- Red-Eye Correction

- Resize

- Filters (Black & White, Sepia, Toy Effect, etc.)

How to Edit a Photo:

- Enter **Playback Mode (▶)**

- Find the photo you want to edit

- Press the **i Button**

- Choose **Retouch** > select the edit you want (Crop, Resize, etc.)

- Press **OK** to save as a **new copy** (your original is preserved)

Quick Fix Tip: *Use Quick Retouch to instantly improve brightness and contrast.*

No-Risk Editing: *Every edit creates a copy, so you'll never lose the original image.*

Favorite and Protect Features

You can mark images you love or protect them from accidental deletion.

To Favorite/Rate a Photo:

- Press **i** during Playback
- Choose **Rating**
- Use ★★★★★ scale to tag your best shots

To Protect a Photo from Deletion:

- Press **i,** select **Protect**
- This locks the file so it can't be deleted from the camera—even by mistake

Organizing Tip: Use star ratings to tag photos you want to print, post, or edit later on your computer.

Transferring Photos via USB, Wi-Fi,

or SD Card

Once you've captured and reviewed your shots, it's time **to send them to your phone, computer,** or **printer.**

You've got **three easy ways** to transfer files:

1. Using a USB Cable

- Plug the USB-C cable into your camera and connect to your computer.

- Your camera will appear like a drive, drag and drop your files.

*Tip: Copy from the **DCIM folder** on your memory card.*

2. Using a Card Reader or SD Slot

- Eject the memory card from the camera

- Insert it into your computer's SD card slot or card reader

- Open the folder and copy photos

*This method is **fastest** for transferring large files in bulk.*

3. Using SnapBridge App (Wi-Fi/Bluetooth)

- Download **SnapBridge** from your phone's app store

- Pair your Z6 III via **Bluetooth** and **Wi-Fi**

- Select photos on the camera and send them directly to your phone

Best for: Quick sharing to Instagram, email, or cloud backup

Setup Path: **Menu** > **Network Settings** > **Connect to Smart Device**

Organizing Tip for New Users

Create folders on your computer by **date**, **event**, or **subject**:

- ▦ 2025-06-20_Grandkids_Visit

- 🔎 Italy_Trip_Landscapes
- 🐶 Max_The_Dog_2025

This makes it easier to find your favorite shots later and ensures your digital memories don't vanish into chaos.

Wrap-Up: Your Camera Is Also Your Digital Darkroom

You don't need fancy software to enjoy or manage your photos. The Nikon Z6 III gives you all the tools to:

- View what you captured
- Improve it instantly
- And share it with the world—without ever leaving the camera

You're not just taking photos anymore. You're curating memories—and now, you're doing it like a pro.

Chapter 12

Nikon Z6 III Hidden Features & Pro Tricks

Your Nikon Z6 III is more than a camera. It's a creative powerhouse filled with professional-grade tools most of which are hidden just beneath the surface.

Whether you want to shoot time-lapse videos of sunsets, record smooth slow motion, or prepare your camera for a silent ceremony or live event, this chapter unlocks the advanced features that make the Z6 III so special.

The best part? You don't need to be a techie to use them.

How to Create Time-Lapse & Slow-Motion Video

Time-Lapse Photography

Capture a series of still photos at intervals, automatically stitched into a video.

How to Use:

1. Go to **Menu** > **Photo Shooting Menu** > **Time-Lapse Movie**

2. Set **Interval** (e.g., 5 seconds between shots)

3. Set **Shooting Time** (e.g., 25 minutes)

4. Choose **Exposure Smoothing ON** to avoid flicker

5. Press **OK**, then start shooting

Great For: Sunsets, clouds, traffic, construction progress

Slow-Motion Video

Record at high frame rates, then play it back slower than real time.

How to Use:

1. Switch to **Movie Mode**

2. Menu > **Movie Shooting Menu > Frame Size/Frame Rate**

3. Select **1920x1080 / 120p**

4. Record video (you'll slow it down in editing software)

Tip: Use a tripod and wide lens for smoothest results

Using Focus Peaking & Zebra Stripes

Focus Peaking

See what's in focus visually—while manually focusing.

Outlines sharp edges in **red, yellow, blue, or white**

Great for portraits, products, or video

How to Enable:

1. Menu > **Custom Settings > d10 Focus Peaking**

2. Choose color and intensity

3. Switch lens to **MF (Manual Focus)** and rotate the ring

Tip: Use on a tripod for precise studio or macro shots

Zebra Stripes

Warns you if part of the image is too bright (overexposed)

- Displays moving stripes over overblown areas
- Helps protect highlights in skin, sky, or white clothes

How to Enable:

1. Menu > **Movie Shooting Menu > Zebra Display**

2. Set Zebra Pattern: Choose **70–100+%**

3. Enable during video recording

*Use **80–85%** for skin tones and **100%** for full exposure alerts.*

Silent Shooting for Events & Ceremonies

Weddings. Church services. Wildlife.

Sometimes, the sound of a shutter can ruin the moment.

Silent Shooting makes your Z6 III completely noiseless.

How to Enable:

1. Menu > Photo Shooting Menu > **Silent Photography**
2. Turn ON

Best when used with natural light (artificial flicker can cause banding)

Bonus Tip: Combine with AF-S and Touch Shutter OFF for maximum discretion.

Custom Shooting Banks (Presets

for Different Scenarios)

You can save your favorite settings into custom banks great for quickly switching between portraits, landscapes, video setups, etc.

Create Custom Banks:

1. Go to **Menu > Setup Menu > Save User Settings**
2. Choose **U1 / U2 / U3** from the Mode Dial
3. Adjust all settings for a specific shooting scenario
4. Save current configuration to that bank

Ideas:

- **U1** – Portraits (f/2.8, Eye AF, natural light)
- **U2** – Sports (1/1000s, AF-C, burst mode)
- **U3** – Video (4K/30p, Face Detection, mic ON)

Now, just twist the dial to switch workflows instantly.

HIDDEN FEATURES & PRO TRICKS

TIME-LAPSE MOVIE
Creates a sped-up video from still images shot at intervals

FOCUS PEAKING
Highlights areas in focus in manual focus mode

ZEBRA PATTERN
Detects overexposed areas by displaying stripes

SILENT PHOTOGRAPHY
Disables shutter noise when shooting

CUSTOM SHOOTING BANKS
Saves settings for different types of photography

Nikon Z6 III

Firmware Updates & Camera Health Check

Keeping your Z6 III updated ensures:

- Improved autofocus performance

- New features and bug fixes

137

- Enhanced compatibility with new lenses and accessories

How to Check for Updates:

1. Visit Nikon's official support site
2. Download the latest **Z6 III firmware**
3. Copy it to your **formatted memory card**
4. Insert into camera, go to **Setup Menu > Firmware Version**
5. Follow prompts to update

Check firmware every 3–4 months, especially before major shoots

Wrap-Up: Turn Your Camera Into a Creative Toolbox

These features may be hidden from the average user but you're no longer average.

Now you know:

- How to craft cinematic time-lapses and smooth slow motion

- How to manually control focus like a pro

- How to disappear into the background with silent shooting

- And how to future-proof your camera with smart updates

Creativity doesn't always come from complexity, it comes from control. And now, you've got both in the palm of your hand.

Chapter 13

Transferring & Sharing Your Photos Easily

Taking a great photo is one thing. Getting it off your camera and into the hands of friends, family, or your favorite social platform? That's where many new users feel lost.

Thankfully, the Nikon Z6 III makes sharing and transferring your images easier than ever. Whether you want to send a favorite shot to your phone, back up a folder to the cloud, or print a framed picture for your wall, this chapter walks you through **every major transfer and sharing method** step-by-step.

Using the SnapBridge App

(Bluetooth/Wi-Fi Transfer)

SnapBridge is Nikon's official app for wirelessly transferring images from your camera to a smartphone or tablet.

How to Set Up SnapBridge:

1. Download **SnapBridge** from the **App Store** (iPhone) or **Google Play** (Android)

2. Turn on your Z6 III and go to:

   Menu > Setup Menu > Connect to Smart Device

3. Enable **Bluetooth & Wi-Fi**

4. Follow on-screen instructions to pair the camera with your phone

5. Open SnapBridge and allow permissions

Once connected, you can:

- View thumbnails of your photos

- Instantly download selected shots to your phone

- Enable **Auto Transfer** for real-time syncing

Great for: Social media posting, texting pictures to family, and backups on the go

Connecting to Your Phone or Tablet (Step-by-Step)

Once paired with SnapBridge, here's how to send images quickly:

To Send Selected Photos:

1. Open SnapBridge on your phone

2. Tap **Download pictures**

3. Select your Z6 III device

4. Choose the images you want

5. Tap **Download** and they'll appear in your gallery

To Enable Auto Transfer:

- In SnapBridge, tap **Auto Download** to ON

- Your camera will send photos in the background as you shoot (JPEG only)

Note: RAW files are large and don't auto-transfer. Use JPEG for mobile sharing.

Quick Cloud Upload Options

After you get your images onto a phone or computer, you can back them up to the cloud to ensure they're never lost.

Best Cloud Options:

- **Google Photos** – Great for Android users and unlimited JPEG backup with compression

- **Apple iCloud** – Seamless for iPhone users

- **Dropbox** – Easy drag-and-drop interface

- **Amazon Photos** – Free full-resolution photo backup with Prime membership

143

*Tip: Use cloud storage for both **archiving** and **easy sharing** links to friends or family.*

Printing Photos from Camera to Home Printer

Yes! you can still print physical photos. And yes, the Z6 III supports direct USB printing and card-based transfers.

How to Print Your Photos:

1. Save your images to **SD card** in JPEG format
2. Insert the card into your home **photo printer** (or computer)
3. Use the printer's LCD screen or software to select, crop, and print

For best results:

- Choose **Glossy or Matte photo paper**
- Print at **300 DPI resolution**

- Set your image to **Actual Size** or **Fit to Frame** depending on your paper size

Use 4x6", 5x7", or 8x10" formats for personal framing and gifts.

Best Formats: RAW vs JPEG Explained (Made Simple)

JPEG (Recommended for Most Users)

- Smaller file size
- Ready to share or print instantly
- Slightly less flexible for editing

RAW

- Uncompressed, full-detail image
- Ideal for editing and professional post-processing
- Not viewable on phones without special apps

Use JPEG when...	Use RAW when...
You want to post to social media	You want to edit lighting/exposure later
You're shooting events or travel	You're doing portraits or product photos
You prefer simple file management	You use Lightroom or Photoshop

*Best Practice: Use **RAW** + **JPEG** mode if you want the best of both worlds.*

To enable:

Menu > Image Quality > RAW + JPEG Fine

Bonus Tip: Use Folders to Stay Organized

Create folders on your SD card or external hard drive labeled by date, event, or subject:

▦ 2025-06-23_Graduation

🏙 Rome_Trip_Sunset

🐶 Bella_The_Puppy

👤👤👤👤 Family_Album_July

Organized files = faster transfers, safer backups, and better printing

Wrap-Up: Take Control of Your Photos

You've done the creative work, now it's time to protect, enjoy, and share it.

Your Nikon Z6 III gives you all the tools to:

- Transfer images effortlessly to your phone

- Back up your memories to the cloud

- Print your favorites at home

- And understand the right file formats for your goals

A photograph is a memory preserved—but only if it's shared or saved. Now, you can do both with ease.

Chapter 14

Maintenance, Care & Safety for Longevity

A camera isn't just a tool, it's an investment, a companion, and a memory-maker.

To keep your Nikon Z6 III working perfectly for years to come, you don't need complicated routines or technician-level knowledge. You just need to know a few simple habits that prevent damage, protect your images, and extend the life of your camera, battery, and accessories.

This chapter gives you the practical steps to keep your Z6 III in top condition at home and on the go.

Cleaning Your Camera and Sensor

Safely

Cameras don't like dirt, dust, moisture, or oily fingers. Regular cleaning keeps your images sharp and your sensor healthy.

Exterior Cleaning

- Use a **soft microfiber cloth** to wipe the camera body
- Clean the LCD gently with a screen-safe wipe
- Use a blower (not canned air) to remove dust from buttons or the viewfinder

Lens Cleaning

- Use a **blower** to remove surface dust
- Lightly brush with a **lens cleaning brush**
- Wipe gently with a **lens tissue** or microfiber cloth in circular motions
- Only use **lens cleaning fluid** when needed

Sensor Cleaning (Be Cautious!)

- First, use the **camera's internal sensor** cleaning feature:

    - Menu > Setup > **Clean Image Sensor > Clean Now**

    - For manual cleaning, use a sensor cleaning swab only if you're confident

    - Otherwise, visit a camera shop for professional cleaning

Tip: Always change lenses with the camera facing downward to reduce dust falling into the sensor.

Battery Care and Storage Tips

Batteries are your power supply but they degrade if mistreated.

How to Extend Battery Life:

- Use only official **Nikon EN-EL15c** batteries or trusted third-party equivalents
- Don't overcharge—unplug after full charge
- Avoid full drains; recharge when battery hits 20–30%
- Store batteries in a cool, dry place, ideally in a padded pouch
- Remove battery from camera when not used for more than 2 weeks

Avoid:

- Extreme temperatures (hot car, freezing weather)
- Charging via unofficial cables or damaged ports

*Keep a **spare battery** fully charged and in your camera bag—it can save a shoot.*

Best SD Cards & Storage Habits

The Nikon Z6 III uses **CFexpress Type B** or **XQD** cards (preferred), but can also work with SD cards using an adapter.

Recommended Specs:

- Minimum 64GB for regular shooting

- For video, use **300MB/s+ write speed**

- Trusted brands: **SanDisk Extreme Pro, Sony Tough, Lexar Professional**

Storage Best Practices:

- **Format cards in the camera**, not on your computer

- Use **multiple cards** rather than one giant one (less risk of total loss)

- Avoid removing cards while the camera is on

- Don't fill a card to 100%—leave 5–10% free

- Use a **hard case** to store SD or CF cards safely

How to Avoid Damaging Your Gear

Your camera is tough, but not invincible. A few simple rules can protect it from harm:

Don'ts:

- Don't leave the camera in a hot car or in direct sun for long
- Don't spray cleaning fluid directly onto the lens or screen
- Don't yank cables from ports—gently unplug
- Don't touch the camera sensor with fingers or cotton swabs
- Don't over-tighten tripod mounts or accessories

Dos:

- Always attach your **strap** before carrying
- Use a **rain cover** or waterproof bag in bad weather
- Keep a **lens cap and body cap** on when not in use

- Zip the camera in a padded bag during transport
- Keep **silica gel packs** in your camera bag to control humidity

Think of your camera like a musical instrument: it performs best when it's cared for, handled with respect, and stored properly.

Traveling with Your Z6 III: TSA & Airport Tips

Whether you're heading across the country or around the world, your Nikon should travel safely with you.

Airport & Flight Prep:

- **Carry-on only:** Never check your camera gear— baggage handling is rough
- Pack in a **padded camera bag** with removable inserts

- Bring extra batteries in **carry-on**, not checked bags (TSA rules)
- Place your camera **in a bin** at security checkpoint, just like a laptop

Travel Essentials:

- 2–3 memory cards
- Spare batteries + charger
- Power plug adapters if traveling internationally
- Lens cloth + compact blower
- SD card case + USB reader

*Before a big trip, do **a full gear check**, test shoot, and charge all devices.*

Wrap-Up: Longevity Comes from Love

When you care for your camera, it will care for your memories.

The Nikon Z6 III is a high-end, precision device that deserves gentle, thoughtful maintenance. A little cleaning, safe handling, and smart storage go a long way in keeping it running like new, whether you're shooting next week or ten years from now.

A clean camera is a clear mind. A charged battery is confidence. And now, you're fully equipped to protect both.

Chapter 15

Troubleshooting Common Problems

Even the best cameras hit a bump in the road. and when something goes wrong, it can feel stressful especially if you're not sure what caused it.

But here's the truth: most camera problems have simple solutions and you don't need to be a technician to fix them.

In this chapter, we'll walk through the most common issues Z6 III users face and exactly how to resolve them, one step at a time.

Camera Won't Turn On or Charge

It's the most unsettling issue: you press the power button... and nothing happens.

Step-by-Step Fix:

1. Check the Battery

- Remove it and reinsert it fully.

- If the battery feels loose or clicks oddly, it might not be seated properly.

- Try a fully charged second battery if available.

2. Confirm It's Charged

- Plug the charger in and look for the orange LED.

- No light = charger not working or outlet issue.

- Light flashing = possible battery fault.

3. Inspect the Power Switch

- Slide it fully to the ON position—some users accidentally leave it halfway.

4. Try Without Memory Card or Lens

- Remove memory card and lens, then turn on the camera.

- A faulty card or lens can sometimes interrupt boot-up.

If nothing works, try a hard reset (remove battery, wait 30 seconds, reinsert). If the issue continues, test with a different charger and battery before contacting Nikon support.

Autofocus Not Working

Your subject is in frame, you half-press the shutter... but the focus won't lock.

Causes & Fixes:

1. Lens is in Manual Focus

- Switch lens to **AF (Auto Focus).** It's usually a physical switch on the lens.

2. Focus Mode Mismatch

- Go to Menu > AF Mode

- Try **AF-A** (Auto Servo) for general use

- **AF-S** = good for still subjects

- **AF-C** = for motion

3. Low Light or Low Contrast

- Point at a high-contrast edge (like where shirt meets background)

- Use the **AF Assist Lamp** for help in dim rooms

4. Face/Eye Detection is OFF

- Turn ON: Menu > AF > Face/Eye Detection AF

5. Dirty Lens or Sensor

- Clean lens gently (see Chapter 14)

- Sensor dirt can confuse the AF system

Bonus Fix: Turn camera off, remove lens, clean contacts (gold rings), reattach lens.

Images Too Dark, Too Bright, or

Blurry

You press the shutter and—ugh—the picture doesn't look right. Here's how to fix each common outcome:

Too Dark

- Increase ISO (try 800–3200)

- Open aperture wider (lower f-number)

- Slow down shutter (1/60s or slower)

- Use Exposure Compensation: tap the **+/- button**, dial to +1 or +2

Too Bright

- Lower ISO (100–200)

- Raise f-number (f/8 or higher)

- Use a faster shutter speed (1/500s or more)

- Use Exposure Compensation: −1 or −2

Blurry

- Raise shutter speed (1/250s or faster)

- Use **AF-C** mode for moving subjects

- Steady your hands or use a tripod

- Tap screen to set focus point manually

Tip: Use Playback Zoom to check sharpness immediately after shooting.

Error Messages Decoded

Sometimes your camera flashes cryptic error codes. Here's what they mean and how to fix them:

Error	Meaning	Fix
"No lens attached"	Lens not seated properly	Remove and reattach lens securely
"Cannot write	Corrupted or	Format card or

to card"	locked memory card	unlock card switch
"Battery exhausted"	Dead battery	Charge or replace battery
"This card cannot be used"	Wrong type or speed	Use approved CFexpress/XQD card
"Image cannot be displayed"	RAW or corrupted file	View on computer or shoot JPEG
Freezing or lag	Glitch from settings or firmware	Turn off, remove battery, reboot
Overheating icon	Extended video use	Turn off and let cool for 10 mins

Resetting Your Nikon to Factory

163

Settings

When all else fails or if you just want a clean slate you can reset your Z6 III to its default factory settings.

Two Ways to Reset:

Option 1: Full Reset

- Menu > Setup > **Reset All Settings**
- This resets shooting modes, menus, custom buttons

Option 2: Two-Button Reset (Quick)

- Hold down **QUAL** + **ISO** buttons together (marked with green dot)
- Hold for 2–3 seconds until the screen blinks

*Resetting erases preferences, **but not photos** on your memory card.*

Wrap-Up: Calm Hands, Clear Mind, Simple Fixes

Your Nikon Z6 III is smart, reliable and like all tech,

sometimes it needs a nudge in the right direction. Most problems are easy to fix once you understand what's going on.

When in doubt:

- Check battery and lens contacts
- Use Playback zoom to troubleshoot focus
- Switch to Auto Mode and test again
- Try a reset to remove hidden conflicts

And most importantly: don't let a small issue stop you from creating. Every great photographer has had a moment where something didn't work and they came back better for it.

You've got this. The camera is in your hands, the knowledge is in your head and you're more than ready to keep creating.

Bonus Chapter

Photography Terms Glossary (Simple English)

A-Z of Nikon Camera & Photography Terms in Clear, Simple Language

📷 **A**

- **Aperture**

The size of the hole in the lens that lets in light.

→ **Lower f-number = more light & blurry background** (f/2.8)

→ **Higher f-number = less light & more in focus** (f/11)

- **Auto Mode**

The camera picks the best settings for you—great for beginners.

- **AF (Autofocus)**

The camera focuses automatically so your subject looks sharp.

- **AF-S / AF-C / AF-A**

- **AF-S:** Focus once (good for still objects)

- **AF-C:** Keeps focusing (great for moving things)

- **AF-A:** Automatically picks AF-S or AF-C

📷 **B**

- **Back-Button Focus**

Focus using a button on the back of the camera (not the shutter button). Preferred by some pros.

- **Bokeh**

The blurry background behind your subject, usually created with wide apertures like f/1.8.

- **Burst Mode / Continuous Shooting**

Takes lots of photos quickly when you hold the shutter

down—great for action.

📷 C

- **CFexpress/XQD Card**

A super-fast memory card used in the Z6 III. Faster than SD cards.

- **Composition**

How you arrange the things in your photo. Good composition makes photos more interesting.

- **Crop**

Cutting out part of a photo to focus on what matters.

📷 D

- **Depth of Field**

How much of your photo is in focus.

→ **Shallow depth = blurry background**

→ **Deep depth = everything in focus**

- **Dynamic Range**

The amount of detail your photo keeps in dark AND bright areas.

📷 **E**

- **Exposure**

How light or dark your photo is. Controlled by ISO, shutter speed, and aperture.

- **Exposure Compensation (+/−)**

Quickly make your photo brighter (+) or darker (−) without going full manual.

📷 **F**

- **Face/Eye Detection**

The camera finds and focuses on people's eyes automatically—super helpful for portraits.

- **Focal Length**

The number on your lens (like 50mm or 200mm). It affects how "zoomed in" you are.

- **Focus Peaking**

A tool that highlights what's in focus when using manual focus.

- **FPS (Frames Per Second)**

How many pictures or video frames your camera can shoot in one second. Higher = smoother video.

📷 **G**

- **Grid Display**

Lines on the screen to help you align your photo using the Rule of Thirds.

📷 **H**

- **Histogram**

A graph that shows if your photo is too dark, too bright, or just right. (Optional for beginners!)

📷 I

- **ISO**

How sensitive the camera is to light.

→ **Low ISO = clean image** (100 – 400)

→ **High ISO = brighter but more grainy** (1600+)

📷 J

- **JPEG**

A common photo file format. Easy to share and takes less space than RAW.

📷 K

- **Kit Lens**

A basic lens that comes with the camera—usually 24–70mm. Good for starting out.

📷 L

- **Live View**

Using the LCD screen instead of the viewfinder to take pictures.

- **Lens Hood**

A plastic ring that attaches to the front of the lens to block sun glare.

📷 **M**

- **Manual Mode (M)**

You control everything: shutter speed, aperture, and ISO.

- **Metering**

How the camera measures light in your scene.

→   Matrix (whole scene), Center-weighted, or Spot metering.

📷 **N**

- **Noise**

Grain or speckles in photos—usually happens at high ISO

in low light.

📷 O

- **Optical Zoom**

True zoom using the lens (better quality than digital zoom).

📷 P

- **Prime Lens**

A lens with one fixed zoom level (like 50mm). Often sharper and better in low light.

- **Program Mode (P)**

The camera picks shutter/aperture, but you can still adjust ISO and other settings.

📷 Q

- **Quick Menu (i Menu)**

Your shortcut menu. Tap the "i" button to change settings

quickly.

📷 R

- **RAW File**

A photo file with all the image data—best for editing, but large and uncompressed.

- **Rule of Thirds**

A photo composition technique: divide your frame into 3 parts and place your subject off-center.

📷 S

- **Shutter Speed**

How fast the camera takes the photo.

→ **Fast = freeze motion** (1/1000s)

→ **Slow = motion blur** (1/30s)

- **Silent Shooting**

Take photos with no shutter sound—great for weddings or

wildlife.

📷 **T**

- Time-Lapse

A video made from many photos taken over time.

Touchscreen Focus

Tap the screen to focus on a specific spot—like a phone.

📷 **U**

- **U1, U2, U3 (User Modes)**

Preset banks to save your favorite settings for quick access.

📷 **V**

- **Viewfinder (EVF)**

The eyepiece you look through instead of the screen. Works better in bright light.

📷 **W**

- **White Balance (WB)**

Makes colors look natural. Auto WB usually works well, but you can adjust if colors look odd.

- **Wide-Angle Lens**

Shows more of the scene—great for landscapes or group shots.

📷 Z

- **Z-Mount**

Nikon's new lens system for mirrorless cameras like the Z6 III. Better performance, faster autofocus.

Wrap-Up: Jargon, Decoded

Now when you see "AF-C with Face Detection and ISO 800 in Aperture Priority Mode," it won't sound like a foreign language—it will make sense.

This glossary is your go-to decoder anytime you forget a term or feel lost in a menu. Keep it bookmarked, printed, or highlighted for quick reference.

📷 Remember: It's not about memorizing everything. It's about understanding what you need, when you need it and growing from there.

Acknowledgement

To every aspiring photographer and filmmaker who dares to pick up a camera and tell a story, this book is for you. Special thanks to my family and friends for their encouragement, and to the creative community whose passion inspires me daily. Your support made this guide possible.